Jürgen Stausberg

Truck route planning and fleet digitization with telematics

FSC
www.fsc.org
MIX
Papier aus ver-
antwortungsvollen
Quellen
Paper from
responsible sources
FSC® C105338

Truck route planning and fleet digitization with telematics

Jürgen Stausberg

How to use cloud services to optimize truck fleets and reduce costs by digitizing processes.

Bibliographic information of the German National Library:
The German National Library lists this
publication in the German National Bibliography;
Detailed bibliographic data is available on the Internet
can be accessed via http://dnb.dnb.de.

The automated analysis of the work in order to obtain
information, in particular about patterns, trends and
correlations in accordance with §44b UrhG ("text and data
mining")
is prohibited.

Publisher:
BoD · Books on Demand GmbH, Überseering 33,
22297 Hamburg, bod@bod.de
Print:
Libri Plureos GmbH, Friedensallee 273, 22763 Hamburg

ISBN: 978-3-7693-0222-6

Table of Contents

2025

How to use cloud services to optimize truck fleets and reduce costs by digitizing processes.

Jürgen Stausberg

WHY THIS BOOK?

Companies with a fleet of vehicles, e.g. from the brewery, bakery, mill, construction logistics, building materials trade, recycling, beverage wholesale, fresh produce service, sanitary wholesale and textile service sectors, often ask themselves where costs can be saved in the fleet and/or where performance can be increased.

This book describes a way to identify cost-cutting potential in the vehicle fleet.

Truck route planning with telematics are the key elements for achieving efficiency gains in the fleet.

Companies that regularly supply customers often have fixed routes. Particularly in times of fluctuating delivery volumes due to the economy, pandemic, weather conditions, etc., fixed routes inevitably lead to empty runs by trucks. Drivers often only know their route and with fewer loads, the costs per transport unit increase for the same mileage.

This book is about counteracting this development. It shows how modern digitalization methods can be used to manage a fleet more efficiently and reduce costs.

Route optimization with telematics helps to find starting points for improvements in your own company's fleet and to eliminate weak points. Modern systems and service solutions from the field of telematics are presented that can help with this.

The book is aimed at fleet managers from different sectors. Not all of the starting points presented are relevant for every industry. Beverage wholesalers will have a different focus than building materials wholesalers, for example. Telematics offers different components depending on the industry.

Route planning

If you ask Wikipedia for the definition of Route planning, you get the following answer: Tour planning is a
planning process in which (transport) orders are combined into tours and put into a sequence. A tour is usually carried out by one person or one vehicle. This planning process is important wherever a large number of
orders and tours have to be planned. Examples include
the delivery of goods to a retailer's branches, the collection of post, refuse collection, passenger transportation and the deployment of service personnel.

The aim of route planning is, for example, to minimize
the number of vehicles used, the distance covered, the
operating time, CO2 emissions or a more complex cost
function. In the standard route planning problem, all
starting and destination points are located in a depot
where a limited number of identical vehicles with limited capacity are usually available. Other variants take
into account additional restrictions such as time windows, several depots or arbitrary start and destination
points. (Wikipedia, Route Planning 2025)

Telematics

If you look at Wikipedia, the first sentence you will find is: Telematics (composed of telecommunications and information technology) is a technology that combines the fields of telecommunications and information technology.

Telematics is a broad term that is used in this book in connection with vehicle telematics.

In relation to the truck, this simply means IT, i.e. the processing of data relating to the truck and the transmission of data from and to the truck. This includes technical data from the vehicle, such as engine speed and diesel consumption, as well as position data and downtimes of the truck or order data with a connection to the navigation system.

The beginnings date back to the 1980s, when special on-board computers recorded the wheel revolutions and thus the kilometers driven and idle times. Special data cards, similar to today's driver cards - or even a cable connection from the truck to an electronic filling station - transmitted the data to the control center for evaluation.

Outdated? Or not? Because not much has changed in terms of data content and its use. But the technology is smaller, simpler and data transmission has been revolutionized with LTE, 4G, 5G, satellites, etc.
GPS positioning for localizing and identifying the

customer's stop or current driving position and the digital speedometer with its various connection options have also been added.

Today, such systems are often installed in 30 to 60 minutes or can be operated directly via the cigarette lighter without installation. In the past, installation took 4 to 8 hours. (Wikipedia, Telematics 2025)

Digitization

The best way to understand digitization in the vehicle fleet is to treat everything that was previously recorded in analog form on paper and then often copied as an electronic version.
Instead of on paper, the delivery bill is kept digitally on a tablet. (Wikipedia, Digitization 2025)

Sensors

A sensor is a technical component that measures certain physical or chemical properties (physical e.g. amount of heat, temperature, humidity, pressure, sound field variables, brightness, acceleration). The measured values are converted into electronic signals that can be processed further.

The following sensors are of particular importance for transport logistics: detection sensors, which today often work with Bluetooth and determine whether an object, e.g. a trailer, is in the vicinity of the truck. Temperature sensors that detect and transmit chamber temperatures on the loading area and fill level sensors that can determine when a container needs to be emptied, particularly in the case of waste containers. (Wikipedia, Sensors 2025)

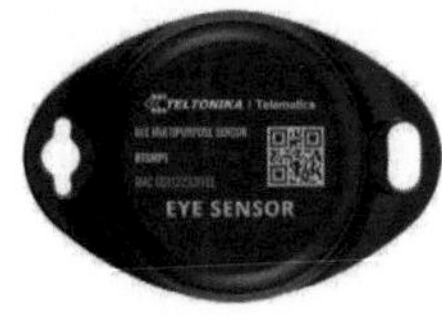

Figure 1   Example of a temperature sensor

The framework conditions have changed

In the past, customers ordered on a fixed date and always ordered approximately the same quantities. The scheduling department assigned customers a delivery day, a standard truck and a standard driver. Truck navigation did not yet exist. The standard driver was instructed for the tour. Only he knew the customer's delivery conditions.

Today, customers are affected by strongly fluctuating customer groups, whether in restaurants due to the pandemic or VAT, or in canteens where many employees are suddenly working from home.

Fixed framework tours immediately lead to severe underutilization of individual trucks and thus to sharply rising costs per transport unit. Cloud solutions make it possible to optimize routes online. Dynamic route planning can optimally compile the routes on a daily basis and transfer the orders to the tablets in the vehicle.

The navigation system also guides drivers who have never been to a customer's premises to their destination, and photos of the unloading point show them how they should behave.

Figure 2   View of a route planning result

 High bandwidths enable the fast and cost-effective transmission of large amounts of data from and to the truck. While location data usually only comprises a few kilobytes per transmission event, order data, signed PDF documents or photos can quickly add up to several megabytes.

If the customer is not present, a photo of the delivery location is taken using modern technology. If they are present, they sign a receipt on the tablet and an un-signed delivery bill is sent to the server as a PDF. A few minutes later, the delivery bill is in the dispatch depart-ment and billing can begin. No more detours to get the delivery bill to the scheduling department. Overtime in the late evening hours for invoicing staff is reduced.

Today, modern trucks have a standardized interface that provides essential data for fleet management and the connection of telematics systems. This is the FMS connector, which is installed in the fuse box or behind the digital tachograph, depending on the type of truck. It provides power, ignition and groundas well as CAN high and CAN low. Data such as diesel consumption, mileage, engine speed, opening status, axle loads and much more is then made available via the CAN bus.

Figure 3   View of the truck interface (FMS connector)

Tacho data can be retrieved

Via the FMS interface and the D8 connector on the digital tachograph, the mass and driver card data can then be sent to the server at the legally prescribed intervals, even if the company card is registered centrally. The shifts can be evaluated and the kilometers driven, driving and standing times as well as violations of the prescribed driving and rest times can be displayed daily.

More powerful sensors

New technologies such as Speed of Light with Narrow Band IOT and new Bluetooth protocols enable improved detection of the fill levels of glass containers (enables route planning only for full containers) and transmission even in environments where voice communication via mobile radio is not possible.

Figure 4View of a fill level sensor for

Some government expects its companies to reduce their CO_2 emissions and provide evidence of this. These reports must be submitted annually and in full. However, companies should not be content with mere status quo reporting but should focus on a data-based sustainability strategy from the outset.
So the question is: how does a company obtain all the data and how much staff is needed to collect and process it?
In the truck, the diesel consumption and CO_2 emissions of the individual journeys can be derived directly from the CAN bus and the optimization calculations using FMS evaluations.

INDUSTRY-SPECIFIC REQUIREMENTS

Every industry has its own questions about digitalization. What they all have in common is the tracking of trucks. The remote retrieval of tachograph data (remote download), the electronic recording of working hours using chip cards.

Here are some examples of industry requirements:

Food/ fresh food service

For the food service is particularly important to note that both fresh and frozen products are delivered by truck at the same time. Different refrigeration chambers must be available for this, so the delivery temperatures are between +5 and -21 degrees and must be documented, often directly upon delivery by means of a receipt printout or on an electronic delivery note. electronic delivery bill. The target group, such as restaurants and canteens, is demanding and requires a fast and flexible response to orders, which are often received during the night, meaning that orders still have to be picked the next morning, which can change the entire delivery schedule.

As a rule, the vehicles are only driven by one driver who has been instructed for the tour and only he knows the unloading conditions at the customer's premises, which hinders the flexible deployment of drivers. With fixed framework tours and fluctuating delivery quantities, individual tours are not fully utilized.

As a rule, the field service has agreed time slots that are often not regularly checked to ensure they are up to date, meaning that truck drivers are often under time pressure to meet certain delivery deadlines.

The goods are usually delivered in roll containers or E1 crates. These transport units are frequently exchanged and it is necessary to track which transport units are at the customer's premises and need to be returned. This requires comprehensive empties management for roll containers and E1 crates.

The mill fleet ca be separated into vehicles for palletized goods and the silo fleet, which delivers the flour in chambers to transport different types of flour and customer batches and pump them directly into the silo at the customer's premises. There are also articulated trucks that transport grain for the mill, among other things.

The challenge lies in the correct utilization of the individual chambers of the silo trucks (different types of flour to different customers) and in meeting the industry's time windows.

It is particularly important to document the unloading conditions at the customer's premises. When pumping the flour from the silo truck into the customer's silo, a full detector must be active to signal when the silo is full. It must be documented that this full detector is actually active and that the customer's hygiene regulations have been observed so that the delivery can be clearly documented later in the event of a pest infestation.

Industrial bakery

Baked goods are often delivered in E1 containers. It must be possible to trace which batch was delivered by scanning. The documentation of full and empty containers is important for the empties account. Temperature monitoring is crucial for baked goods that are delivered chilled or frozen, for example. On delivery, the full and empty quantities are documented by the temperature stamp on the electronic delivery bill.

Sanitary wholesale

In many cases, sanitary wholesalers deliver to plumbers in the early hours of the morning, when no one is available to receive the goods. In this case, it is very helpful to document the delivery with photos.

As there are always ad hoc tours, the driver must be shown the correct sequence of deliveries for navigation on the tablet.

Determining the delivery costs per installer is appropriate for the calculation.

Construction logistics

In construction logistics, a lot changes during the course of a day, which means that route changes must always be transferred to the tablets in an up-to-date manner and messages to the drivers should also be sent directly to the tablet in a closed system. Construction logistics calculates on the basis of hours, kilometer flat rates, route flat rates and must keep an eye on toll costs in particular as an important cost factor.

A pre-calculation of the individual tours taking into account the toll and a post-calculation based on the kilometers and times driven is essential for the industry.

Invoicing is always based on a delivery bill signed by the foreman, which is completed by the driver with regard to times, tours, waiting times, etc. Waiting times are part of the invoice and must therefore be signed by the foreman. Waiting times are part of the billing and must therefore be signed by the foreman. Transmission via the mobile phone network enables immediate invoicing while the truck is still on the road.

In beverage wholesaling or in breweries, the pre-loading of trucks is of great importance. The loadmaster must know exactly in which order and on which side of the truck body the goods should be placed.

To do this, the driver must document the important data the day before the tour. In modern companies, this is done on tablets in the vehicle. The forklift trucks are instructed accordingly to bring the goods to the correct loading point, and the loadmaster can document whether he has secured the load.

The topic of empties is particularly important in beverage wholesaling. The driver enters the full and empties quantities on the tablet or receives the full quantities from the server and changes them or enters the empties quantities.

The topic of route optimization is very relevant in the beverage wholesale trade. As the standard routes are often provided with time windows that are constantly changing, electronic route optimization helps to reduce the number of kilometers.

Every stop at the customer is recalculated and used for negotiations with the sales organizations.

The recycling companies often work according to fixed routes, according to which they empty the garbage cans and containers. It is important that the driver knows which streets he has already traveled. Ideally, this is done electronically on a tablet instead of on a city map.

Route optimization can be improved if the fill levels of large bins and containers are known to the server in advance via fill level sensors.

Recycling vehicles often drive through very narrow streets, where side and rear view cameras give the driver a better overview, ideally also through video documentation in the event of damage and thus documentation for the insurance company.

Textile service

The supply of hotels, clinics and restaurants with fresh laundry in roll containers requires the simultaneous collection of soiled laundry.

Due to seasonal or operational fluctuations in the volume of laundry, it is often not known how high the proportion of soiled laundry will be on the return trip. A planned value for the return collection must therefore be stored for route optimization based on historical values and the operations' capacity planning. Modern systems use machine learning and artificial intelligence here.

It is important to document the exact number of roll containers using barcode or RFID systems for incoming and outgoing goods at the customer's premises.

Vehicle transportation

The delivery of new vehicles from the seaport or a production facility to the dealers or the transportation of breakdown vehicles to support points and workshops places high demands on scheduling and documentation. The pick-up and delivery mode in route planning shows the dispatcher which vehicles are to be used for which routes. Tracking gives the dispatcher, but above all the end customer, the certainty of when the vehicle will arrIve (if the arrival time is determined for all stops, taking into account the unloading times and routes). Photo documentation during loading and unloading and the user software on the truck tablet specially configured for vehicle transportation are important.

Fuel trading

Tankers usually have several compartments for different types of fuel. This increases the complexity of scheduling. Route optimization for tankers must therefore take products and compartment sizes into account and thus dynamically generate new routes on a daily basis from the current orders and quantities. It is important that the driver has photos of the unloading situation and navigation instructions on his tablet, regardless of whether he has previously been to the customer or not. It is often important to know exactly when the tanks were filled. The opening and closing of dome lids can be monitored with special ATEX-certified Bluetooth sensors, which then send the opening and closing processes to the telematics box and via this to the server.

There are many anecdotes to report in connection with telematics and digitalization processes. The following are just a few typical ones that are intended to provide an introduction to the starting points for increasing efficiency.

Time window not updated

Rügen, morning delivery, locate the truck and realize that it is zigzagging, but why?

The driver: "I'm driving in circles because the second customer on the route doesn't get the goods until 11:00 a.m.". The customer file is checked: The delivery window was between 11:00 and 12:00, entered by the sales representative many years ago (the sales representative has long since left the company).

The customer asks - why this time slot? "Oh, that was back when I used to walk my dog Pfiffi, but the dog died long time ago, you're welcome to come earlier!" Result: kilometers, CO_2 emissions, toll costs and diesel costs could be reduced.

Figure 5Typical delivery situation beverage industry

Odysseys cleared up

Vehicle tracking is activated in the truck, the vehicle is located on the map in the dispatching system at a position far outside the delivery area.

The dispatcher asks telematics support whether the system is faulty, which is not the case. The driver cannot be reached by phone.

The police use the GPS positions from the telematics system to drive to the destination. The driver is completely confused but still in good health; his disorientation is due to the many roadworks and detours and the fact that he is not using the navigation system. The truck could only be found thanks to online tracking.

Wage injustice

The driver comes to his boss and complains about the injustice in the fleet. The boss is astonished. The driver replies that he shows maximum productivity every day and drives many tours.

Other drivers drive to the filling station after 3 p.m., meet up with others there and then return to the yard punctually at 4 p.m. at the end of the working day without having driven any other routes - for the same pay.

Relief for the driver

The police issued a ticket for a truck that allegedly entered a one-way street at a certain time. The tracking from the telematics clearly shows that the truck was in a different place at that time. The documentation transmitted to the police exonerates the driver and helps to avoid points and fines.

Dummy leads to behavioral change

At the start of a telematics project, the tracking computers should be properly connected by the workshop to power, ignition iginition and ground in the fuse box.

However, the fleet management had already fitted the devices to the windshield without connecting them and informed the drivers that the trucks would be tracked in future. Strangely enough, after this announcement, all vehicles arrived home about an hour earlier the next day without the trucks having been tracked.

Waiting time abused

A driver who works for an hourly wage is waiting at a customer's premises, with two other trucks in front of him and three trucks behind him.

When it's his turn, he let ahead the vehicles behind him, as he is paid by the hour while he waits and can therefore maximize his wages.

Detours into the traffic jam

There is always a traffic jam at a highway junction on Friday afternoons, the truck can drive back to the company via the main road or via the highway junction.

The driver regularly drives into traffic jams to read the newspaper and increase the hours to be billed.

Sales revenue with rising logistics costs

The sales department is constantly expanding the delivery area due to its sales commissions, and the trucks have to cover ever greater distances for delivery.

Nobody in the company notices that the fleet costs are rising, and the costs per transport unit are also rising.

The inefficiency of serving distant customers only becomes clear when the delivery distances per customer and the downtimes at the customer are visualized via the customer result calculation.

Clearing service not included

The entrepreneur wonders why the ratio between costs and gross profit is particularly poor for some customers. The cost-impact analysis shows that the stopping times at the customer are too long. The experience was made that the driver arrives late in the evening for scheduling.

When asked: "Why so late?" the answer was: "I was still stocking the shelves at the customer Meier." This stocking service was neither included in the cost calculation nor in the customer agreement.

Mass data collected in the rain

In the words of one fleet manager: "Now I have to go out again in the rain and snow to read the mass data from the vehicles' tachographs". An unloved and time-consuming task. Although the driver card data is archived, it is rarely actually checked.

Drivers are informed automatically, but the value of the data for controlling purposes is generally not utilized. The fact that the remote tacho download can be automated and is therefore more cost-effective than manual archiving has not yet been taken into consideration.

"Our drivers are like Jehovah's Witnesses," said one works council member. His drivers have lots of extra stops to ring strangers' doors and look for the exact unloading point because they don't know the new development area. (Building materials trade).

If the delivery addresses were visible on a map before delivery, the fleet manager could manually determine the correct delivery location in the case of industrial estates or new development areas and transfer this position data to the vehicle navigation system. The map material in navigation systems is often not up to date, especially in new development areas.

However, if the navigation system knows the destination coordinates, the driver can always get directly to the construction site or the new development area without expensive search times.

"Delivery drivers wanted!" These days, signs like this are stuck to the windshields and tailgates of many transport vehicles and often document the high turnover of drivers and the need to retain them more strongly than before and to develop criteria for motivating drivers more strongly than before.

This brings us to the core issue in the fleet sector: driver motivation! Drivers are often left to their own devices, have little communication with the boss, receive many calls a day from the "theorists" in the dispatching department, are constantly under time pressure and, in the worst case, there is a sudden blackout, the truck collides with other road users and the driver ends up in hospital.

How can such situations be avoided? How can appreciation be increased for one of the most important employees in the company, but one who is at the end of the value chain and receives so little attention?

The driver receives an introduction to the truck, is trained as a co-driver on the tour and is then often left alone. In the evening, the production employee can see exactly what he has achieved and is sometimes told on large display boards how many tons have been produced today and at what reject rate. The driver rarely receives any feedback, perhaps by telephone if the customer calls to find out where the goods are and the scheduling department has to ask.

Before positive or negative feedback is given to drivers, it is important to measure and jointly develop target values. This brings us to two key concepts that form the basis for successful driver motivation: Measurement and target values.

Figure 6   Focus on driver motivation

In 1891, the British scientist William Thomson, also known as Lord Kelvin, said: "If you can measure what you are talking about and then express it in numbers, then you know something about it. Otherwise, your knowledge is very poor and unsatisfactory ... only when your knowledge is based on measurements and figures can you free yourself from making gut decisions, guesses or marginal improvements." (Wikipedia 1891)

The statement is still true today and has been widely quoted, for example in the European Journal of Epidemiology. . Something similar can also be found in the words of management guru Peter Drucker: "If you can't measure it you can't improve it." (Drucker 1954) (Lavinsky 2025)

Examples of reliable measured values in the vehicle fleet:

- Loading times in the yard
- Departure times from the farm
- Utilization of the vehicle on the tour
- Kilometers of the tour
- Service life with customers
- Downtimes to comply with the statutory breaks
- Reaching the customer's time window
- Compliance with temperature limits
- Detection of unauthorized opening and closing of unloading flaps/doors
- Compliance with the target values for the tire pressure
- Number of small claims
- Cleanliness of the vehicle
- Number of suggestions for improvement

Target values

Let's move on to the most important point when looking at the vehicle fleet: the target values.

If you ask fleet managers how high the maximum delivery costs per customer and tonne can be in a certain delivery region, you often get a shrug of the shoulders and the comment: what's important is the bottom line. But in many cases, the bottom line could be more.

The reason for this lies in the lack of training and knowledge of how the costs of an individual customer delivery are made up and how high the target costs per transport unit may be, especially in distribution transport with many customer deliveries on one tour.

Interesting results are presented in the book: The Problem of Realizing Performance-Related Behavior of Drivers in the Company's Own Vehicle Fleet. (Stausberg 2001)

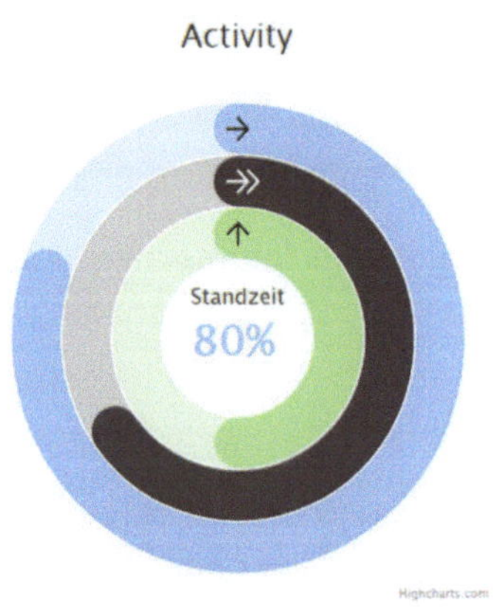

Figure 7Example of target value achievement from the driver's perspective

As early as 1975, Latham described how the company was able to increase vehicle utilization from 60 % to 90 % by setting high targets . In Canada, timber drivers were given particularly large pins (best driver of the week) if the performance agreed in advance was achieved or exceeded. (Latham 1975)

Competition among drivers has often arrived in companies these days when it comes to diesel consumption analyses. However, the holistic view is missing in many cases, even though it is particularly important.

According to Lattmann, the following must be taken into account when setting goals : (Lattmann 1977)

- The goal must be accepted by the employee.
- It must be substantial.
- It must become the subject of communication between employees and superiors.

Achievement motivation

What motivational effects does this have?

Once the overall objectives have been defined, they must be broken down to the level of the drivers.

Examples: Departure time from the yard 6:00 h, loading time per truck 20 minutes, unloading time at the customer according to the target values achieved for fixed unloading (laying the hose) and variable unloading time (pumping time per ton), achievement of the stored time windows, achievement of the planned target km on the tour, tire pressure at 9 bar on the drive axle, etc.

On the other hand, there is the actual performance that needs to be communicated. Important points are (Benston 1972) :

- Knowledge of performance influences learning behavior and the standard of performance that can be achieved through learning.
- Knowledge of performance influences motivation. The best-known phenomena of performance knowledge are increases in motivation.
- The more specialized the knowledge of the performance, the faster the improvement and the higher the performance level.
- The later the performance is announced, the less influence it has on the performance.
- If knowledge of performance decreases, performance decreases.
- Knowing the performance and comparing it with the target or the performance of others creates a performance competition.

The digitalization and optimization process in the fleet is always a matter for the boss. Time must be set aside for this, even if only a little, as most of the work is done by correctly configured systems. The support of the management is required for the implementation of measures and the activation of sub-projects. The sub-projects must be managed by internal project staff.

Carry out controlling and not control

This book is about the advantages of using telematics in the vehicle fleet. Many people understand this to mean that the driver can finally be controlled at all times. You can monitor the truck, where it is and what detours it is taking. Caution: control is demotivating.

Controlling, on the other hand, is understood as a management tool. It leads to improvement targets and motivation through teamwork and that is what this book is about.

To motivate everyone and increase performance in the fleet.

For example, what could the specific goals look like?

We want to reduce the transport costs per transport unit in the fleet by 15 % per year!

Those responsible must comment on target deviations of
10 % (for selected score cards) a comment to the superior including activities!

The sales department must take distance-dependent delivery costs into account in the calculation!

Additional costs due to customer reorders are collected and used for calculations!

We want to manage with x% fewer vehicles or drivers than before for the same transport volume!

We want to save on fuel costs per year!

We want to reduce overtime by 10 % per year!

We want to be able to tell the customer the current delivery time immediately in case of queries!

We want to reduce the kilometers driven and travel times in relation to transport performance by 10 %!

We intend to reduce empty runs by 5%!

We want to reduce the administrative work for driver cards and temperature monitoring by 30 hours per year!

We want to document the delivery times, whether they were within the time window!

We want to document that the temperature at the time of delivery of the chilled goods has corresponded to the standard value!

We want to give the driver feedback on his performance from the previous day.

We want more safety for the driver to avoid accidents - by installing cornering assistants and reversing cameras with AI person recognition.

We want to reduce damage to vehicles by 10 % annually!

Develop optimization approaches

Check routes

Do you do the same? Are similar regular tours driven by the same driver to the same customers in the same delivery districts every day? Does the driver determine the customer sequence for the tour himself?

Then you have probably already discovered the first potential. Take a new driver, equip him with navigation so that he can find the customers, and suddenly you realize that completely different routes are being taken and the driver is back earlier than the standard driver.

There can be many reasons for this. There is the "Truckers Inn", which is regularly visited at lunchtime because of the large portions, which may involve detours, possibly even toll costs and journeys outside the tour.

There is also the driver's desire to generate as many hours as possible with overtime pay. So maybe you'd rather drive into a traffic jam?

Optimize routes

Transfer your orders with delivery address and transport quantity to the electronic route planning and have the order of delivery (order optimization) or a complete, daily dynamic optimization of your delivery area carried out. The results always show the expected planned costs per transport unit on the tour.

Figure 8Dispatch board with orders and tour costs

With complete dynamic route optimization, orders are optimized according to vehicle capacities, delivery time windows and costs independently of existing regular routes. In practice, this step often shows a savings potential of 15% or more.

The optimization results must then be communicated to the scheduling department so that the orders are processed and loaded in the correct order.

At the same time, they are downloaded to tablets. The navigation guides the driver according to the optimization sequence and he can then, for example, record the empties and create the electronic delivery receipt at the customer's premises.

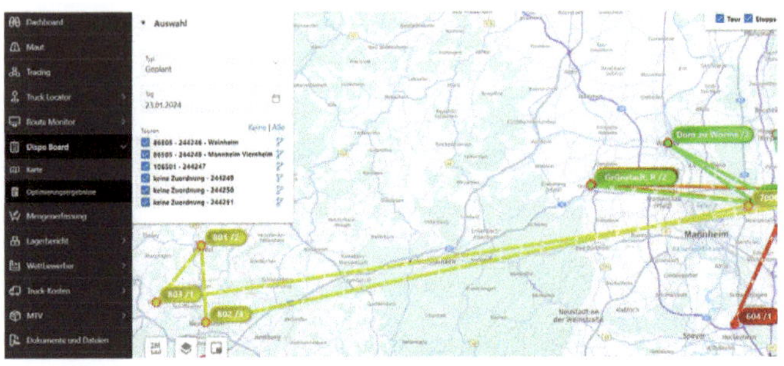

Figure 9Route optimization view

Determine delivery costs per customer

You have acquired new customers, negotiated the conditions and deliveries are made week after week.

Do you know what this delivery really costs?

Record the variable and fixed material and personnel costs and allocate them to the customers according to their origin based on star distances. This can only be done systematically with special geo-functions in a database and the use of telematics.

In order to determine the route and customer costs in line with their origin, the kilometers driven must be evaluated with variable costs and the driving and downtimes with pro rata fixed costs and personnel costs.

The customer addresses of the delivery orders are geo-coded, the items are displayed on the map and the pro rata travel costs and stand costs (direct costs) are calculated based on the customer's location.

For companies with different orders per stop (e.g. 5 customers per stop due to the unloading conditions at the marketplace), an additional order cost calculation can be carried out in which the stop costs are distributed again.

You can then see whether the gross profit and the logistics portion contained therein are covered.

They recognize whether the customer's distance from the depot, the product range or the driver's behavior during unloading or detours are the cause of deviations in the calculation.

Depending on the unloading quantity, the work to be carried out at the customer's premises must be stored with planned times, which lead to planned costs and can later be compared with the actual values of the tour. With the help of the driver and ideally electronic documentation on a tablet, waiting times and incorrect delivery time windows should be documented.

Figure 10Unloading situations at the customer's premises are decisive for the amount of customer delivery costs (lifting platform, roll container exchange, roll container packing, etc.)

Introduce fleet controlling

You may be evaluating the driver cards in a time-consuming and cost-intensive manner and recording the driving and standing times for working time accounting per day and stop.

Why bother if nothing else happens with the data? In view of this, the data is often not collected in the first place.

Fleet controlling of the classic kind is not usually carried out because the data is available in the various accounting systems. The acquisition values of the trucks etc. are stored in the asset accounting, the fuel costs in the operating cost center and the wages in the personnel accounting. As a rule, nothing is considered integrated.

In logistics, the costs caused by idle time, kilometers and travel time must always be seen in relation to performance. And that is the unloading performance measured in terms of transport units (example: number of crates, barrels, roll containers, pallets, load units).

The data from the various accounting systems must be combined. Viewed in isolation, the figures are not very meaningful. Comparative logistics figures are hardly possible without IT support.

In addition, important logistics key figures such as kilometers per stop, downtime per transport unit and truck capacity utilization must be created on a company-specific basis and stored with target values.

Examples:

Weight utilization
Payload of the vehicles in tons in relation to the actual load in tons.

Volume utilization
Loading capacity of the vehicles in transport units in relation to the actual load in transport units (e.g. pallets).

Time utilization
The time utilization of the vehicles based on the target time of e.g. 8 hours per day. If a vehicle is only in use for 4 hours per day, the key figure is
50 %. The base values should be formed on a company-specific basis.

Kilometers per transport unit
Kilometers driven per transport unit on the respective tour. This key figure says something about the quality of the planning. If, for example, more kilometers are covered per transport unit, e.g. pallets, than planned, the vehicle was not properly utilized on average and the routes and delivery sequences were not optimized.

Chamber temperature/stop
In some industries, it is important for the fleet manager to keep a record of temperature compliance during transportation and loading. Existing systems often have the disadvantage that the loggers in the vehicle have to be disconnected manually and then read out for each truck in the fleet office. Temperature alarms that

indicate incorrect temperatures online are not possible in this case.

On request, configurable evaluations offer the additional operational benefit of linking the discharge temperature to the discharge location and promptly indicating any temperature limit violations.

Degree of time (prouctivity)

It is important for the driver to know whether the expected performance has actually been achieved.

Performance can only be controlled if the performance has been measured and discussed.

In some industries, it is possible to define a target time for loading and unloading for each customer type per transport unit. These target times are easier for the driver to handle than the gross yields, etc. He must know what target time is expected of him for the individual customers, whether he has reached it or not.

An important benchmark for the driver is therefore the degree of time. The degree of time Is the quotient of the fixed target time per customer delivery for maneuvering and delivery note processing and the variable, quantity-dependent target time for the individual transport units in relation to the actual time at the customer. Measuring performance in this way enables performance comparisons with other drivers.

Every week, put up a list showing the driver and his vehicle and how fuel consumption has developed.

Bring in a trainer for a few days to familiarize drivers with the latest fuel-saving techniques optimized for the vehicle.

Regular consumption checks after training can be carried out via the truck's CAN bus.

Checking the truck type

Tail lifts and load securing, lifting platforms, fixed body with advertising lettering and tarpaulin body with sliding tarpaulin, trailers and semi-trailers? Have all the options checked and the drivers tested. You can save costs at alone by optimizing the vehicle type in the body.

Carry out driver feedback

The easiest way to realize the cost-cutting potential in the fleet is with motivated drivers.

To achieve this, you need to talk to them, measure their performance and give them daily feedback.

Examples of performance: fast loading and unloading, short downtimes, clean work clothes, clean trucks and much more. Good performance can be measured by targets: Were the goods delivered in the expected time window, was the unloading time reached? Were the correct routes chosen?

Streamline administration

Some companies have the drivers write driving reports, which are then filed because further processing is too time-consuming. Use telematics for electronic trip reports and reduce the manual workload in administration. This also includes automatic, electronic working time recording.

Figure 11Manual driving report still useful?

Reduce phone calls

Check your telephone bill and the costs for communication between the control center and vehicles!

This is less about telephone costs, which today are often covered by a flat rate. Of particular importance is the time spent on phone calls between fleet management and drivers, which is not usually included in any expense report. In addition, there is an increased risk for drivers if they are distracted by phone calls while driving. If the driver receives a new order by phone for which he has to make a note of the unloading point, he has to drive to a rest area. This also costs time. There is also the risk of incorrect journeys and detours due to incomprehensible delivery addresses.

DATA/ACTIVITIES FOR THE OPTIMIZATION PROCESS

Before logistics processes can be improved, there is a long way to go before data can be processed and joint measures discussed and adopted.

Data is needed to determine savings effects. Which areas must the data come from?

Customer data

Use the driver as the most important link to the customer. Drivers often know exactly which order time windows, intervals and minimum order quantities change for the customer.

The drivers can give the fleet and sales management important impulses as to which activities the customer can develop in order to reduce unloading costs, starting with the preparation of empties through unloading assistance.

Unloading location

It is important that the dispatcher defines the unloading location directly on the map.

Telematics, integrated navigation: The telematics unit in the vehicle ensures communication with the server and receives the orders via the mobile network, which it forwards to a navigation solution. The destination positions, longitudes and latitudes are already assigned to these orders, so the driver only has to tap on the order and the route to the destination is planned. Advantage: The driver does not have to enter any addresses. Destination positions, e.g. in new development areas, are already set and corrected by the fleet manager. There is no need for the driver to make enquiries and the associated telephone costs.

Customer type

The customer's type data, consisting of fixed (unloading quantity-independent) and variable (unloading quantity-dependent) times, are of great importance for route optimization. The product groups to be delivered are to be typified and provided with the target time for unloading. Telematics helps to carry out this typification based on historical data.

Photos

Drivers contribute comments and photos of difficult unloading conditions (e.g. long distances with hand trucks) to ensure that the planned values for cost accounting and optimization calculation are set correctly in the system.

Time window

Customers must be served in the right time slots and on the right days of the week. This is the only way to optimize routes.

Distance classes

Distance zones are created around the depot and the target cost rate and the target cost revenue per transport unit are defined for each distance zone per product group. Example: The delivery of beverages within a radius of 35 km around the depot must not cost more than x euros per transport unit.

Target costs

The target costs per transport unit must be stored for each customer according to type in order to carry out an actual comparison.

Transport quantity

If you want to carry out a savings calculation, the costs must always be related to the transportation service. This can be transported crates, barrels, pallets. containers, roll containers, etc. In many industries, these quantities are assigned to orders via the merchandise management system. As a rule, this data is transferred from the ERP system via JSON/residual interfaces.

Alternatively, quantities can also be allocated using special programs, scanners and simple data capture.

Technology

The vehicle's technical data includes the internal number, the registration number, the total weight, the payload, the year of construction and the number of axles, length, width and height of the vehicle, which are important for calculating the toll. Annual mileage.

Costs

The costs can be calculated as annual values or automatically recorded as monthly values from the interfaces to UTA, DKV, Toll Collect etc.

Data from accounting are e.g.

- Purchase value of the vehicle incl. body excluding VAT,
- Taxes and insurance for the vehicle,
- the imputed depreciation period and the imputed interest rate or lease payments.
- The repair costs and fuel costs can be taken from the material cost statement.

The HR department manages all personnel costs. The hours worked by the drivers are recorded for the company per driver and month, with details of overtime.

These data are used to create billing rates that are used for route evaluation and customer billing.

The ID of the driver card is stored for the automated digital-tachograph-data download and the RFID of the driver card or the ID key from the key fob is stored for the working time accounting.

Tour dates

Times and services

To calculate the savings, the following data must be automatically recorded by the vehicle:

Unloading location, customer number, kilometers, driving time, standing time at the customer and standing time not at the customer, standing time in the yard, tour start, tour end.

The tour time extends from the yard exit to the yard
entrance and usually starts at a depot.
depot.

In many cases, loading takes place in the yard before
the tour begins. This can be done with or without the
driver present. The driver's loading activity shortens his
available working time for the delivery time. If working
time recording is provided by telematics with a chip
card, the difference between the start of work and the
departure time of the truck can be determined, including whether the departure check was carried out or
not.

Modern telematics systems report starts and stops via
the ignition signal or a motion sensor. Automatic route
generation is integrated into the systems so that the
routes for each delivery day and each truck are recorded fully automatically and displayed on the map.

However, it is not enough for the tour and its many
stops to be recorded automatically. The tour information must also be available quickly at the control
center, because the fleet manager must be able to talk
to the driver about it after the tour.

Modern systems automatically convert this into a score
card that the driver can see as feedback on their tablet.

Only then are the memories still fresh and the driver
can give important impulses as to what was good and
what needs to be improved.

Logistics key figures

The orders are provided with their quantities from scheduling and optimization and linked to the actual data of the tour after completion. The degree of utilization is calculated from the kilometers and times and finally the costs per transport unit for the individual delivery stop are shown.

This allows driver productivity to be derived.

Evaluations per user with definition of key figures and individual query options for each task area increase user acceptance.

Driving and rest times

Special telematics systems and evaluation modules enable the remote download of mass memory data and driver cards from digital tachographs.

Remaining driving times can be prepared for scheduling during the tour.

The data evaluation takes into account shifts, kilometers and working hours. Violations are documented and the legal archiving obligation is fulfilled.

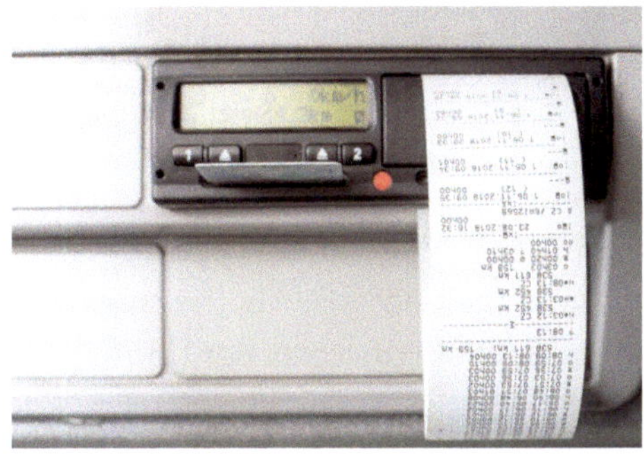

Figure 12Data from the digital tachograph

SELECTED USERS

Controller

Depending on the size of the company, companies may have controllers or commercial managers. Their aim is to maintain or increase the profitability of the company. In wholesale, where warehousing and transportation are the two enormous cost blocks, transportation is often the decisive factor for profit and loss with a customer.

What are the exact costs for individual deliveries to individual customers? To answer this question, it is important to create as large a quantity structure as possible, with meaningful average values and reliable basic data.

A customer cost calculation based on the telematics data shows the cause of their occurrence, the controller sees the target deviation in the costs per transport unit and whether the causes were due to excessive working times or the routes with traffic jams or detours.

Perhaps the customer is also located in a delivery area far outside the normal range for which the normal calculation may not be applied? Special times, for example due to crane unloading or waiting times, may be paid for by the customer if they can be documented. The controller then has to approach the sales department with control.

It may also turn out that the margin for individual products is not sufficient to finance transportation by truck and that it is therefore better to choose parcel shipping.

Workshop manager

Vehicles need to be serviced regularly to minimize downtime. It is therefore important for the workshop to know the current mileage of the vehicles. If the service interval with the required service work is stored in the database, the workshop manager can be notified automatically that a particular vehicle requires servicing.

Personnel department

If working time recording has not yet been automated by appropriate chip card systems, drivers write down their working times. As a rule, the HR department is very interested in receiving the data automatically, and EU regulations also require this.

Now to the remuneration system: As a rule, drivers are paid according to hours and overtime, which means that the slow driver sometimes receives a higher wage than the fast, motivated driver.

According to Section 7 (1) of the German Driving Personnel Act, the remuneration of driving personnel may not depend on the quantity of goods transported or the number of tours completed.

This is because these can have the effect that the driver has an interest in getting to the depot as quickly as possible, i.e. also by driving as fast as possible, and reloading again in order to drive out more volume. This can impair road safety, which is not in the interests of the legislator.

The way out is: Not the speed of driving, but the productivity of the individual unloading process is rewarded.

The degree of time has already been described above as a measure of productivity.

A monetary factor is linked to the average amount of time worked per month, which can lead to a variable component of income.

Furthermore, the fleet manager should also be able to reward individual top performance - from low fuel consumption to clean vehicles, many other components are conceivable.

Sales

Sales are traditionally measured by the achievement of sales targets. In modern companies, gross profit is included in the analysis. If it expands sales in distant delivery areas, which leads to higher distribution costs, the logistics costs incurred can alternatively be considered in the sales commissions.

Distribution must be included in the cost reduction project for logistics, as it is often the cause of incorrect delivery days, incorrect time windows, incorrect routes and underutilized vehicles.

Managing Director

The timeliness of target/actual deviations is important; the company boss usually wants to be able to see at a glance via score cards and graphics where things are getting out of hand. Modern systems have stored who has to react to which deviation and how.

The major advantage of cloud solutions in the cost-cutting process is that everyone involved in the fleet has access to the same data and can also view it at different company locations via the internet. This results in savings in meeting time and travel time. All parties can access the same information, provided they are authorized to do so, and discuss the same facts.

The boss can always call up the degree of target achievement for each individual departmental target. Conversely, the employees know for each tour that they have to explain certain deviations to the boss in case of doubt, which can help to prevent costs from rising in the first place.

What are the efficiency benefits?

The market is complex and confusing. If you enter the word gps-tracking into Google in January 2025, several million search results will come back. How are you supposed to find the right solution for your business?

Since we don't know exactly what we can achieve with telematics and what technology can do, we are unsure and therefore lose the efficiency advantage if we don't act. But how should we proceed?

What options are there?

You want to see where the driver is at the moment and thus also estimate when he will be back at the company yard. In many cases, the current position of the vehicle with the time stamp via a tracking box is sufficient. In other cases, however, arrival monitoring with consideration of the traffic situation is required, in which case a more complex navigation system is needed.

If you want a digital signature at the customer's premises, an MDE device or tablet is required.

Let's move on to the different variant classes, which can only be a rough classification here, because there are (see the Google query) an infinite number of different variants on the market.

Variant class 1: Locating base

In this class, the entire Plug&Play tracking computer is mounted either on the inside of the windshield or directly on the cigarette lighter.

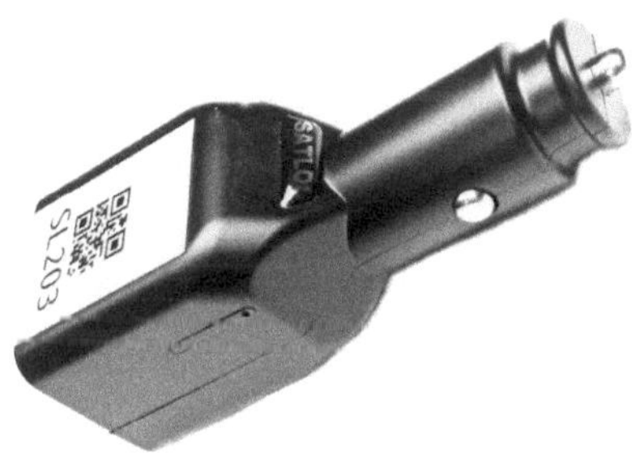

Figure 13GPS tracker for cigarette lighter operation

In addition to the processor and the GPS receiver, these computers have an internal SIM card for data exchange with a central server and send the GPS positions and time stamps as well as kilometers (determined from GPS) cyclically or after an event (e.g. ignition off, ignition on, movement stop).

With variant 1 you can control vehicle positions, you have a low level of complexity, but also only limited benefits. This variant is generally used for proof of concept.

Variant class 2: FMS telematics

"The **fleet management** *system* **(FMS) interface** is a standardized interface to some vehicle data of heavy commercial vehicles. The seven European manufacturers Daimler, MAN, Scania, Volvo, Renault Trucks, DAF Trucks and IVECO joined forces in 2002 to create the so-called FMS standard in order to enable cross-brand traffic telematics applications. (Wikipedia 2025)

There is a standardized 12-pin green telematics connector, which is sometimes fitted as standard or ordered when the vehicle is purchased. Telematics systems designed for this purpose can be connected to it, which then also record and send diesel consumption and engine speeds in addition to the data of variant class 1. In addition to the standardized CAN interface, vehicle manufacturers have their own much more in-depth, primarily technically oriented telematics applications in their product range. Accessories are also available for the FMS device types (which, however, can only be operated on continuous power, ignition and ground):

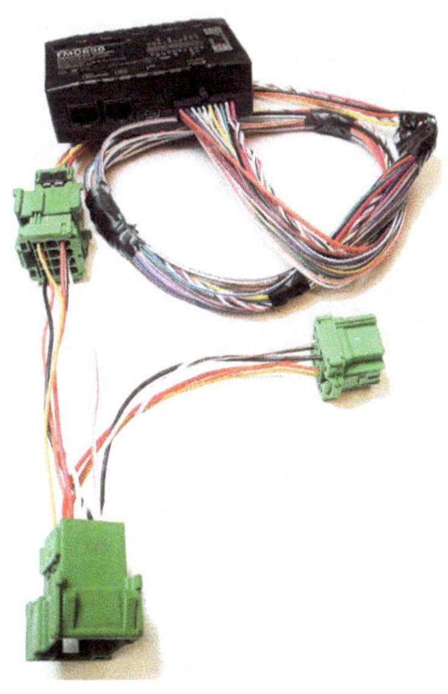

Figure 14Professional GPS tracker with FMS connection cable

1-wire accessories

Driver ID key fobs can be read out with a cable connection. There are special temperature sensors available at , which can also be connected via 1-wire.

RFID accessories

With special RFID readers, active RFID transponders can be evaluated over distances of up to 150 m. Door or flap open/closed, container loaded/unloaded, temperatures in different compartments. This is particularly interesting for quick retrofitting.

Tire pressure monitoring system

Low tire pressure alarms and transmission of information to displays and the control center.

Digital Tachograph connection (with adapters)

Readout of e.g. driver ID, driving time status, break status and transmission of driver card data and mass data.

With variant 2, you can evaluate stops, kilometers, diesel consumption, tour costs, truck costs, vehicle deployment times and make statements on consumption values. The manufacturer-certified plug can reduce installation costs.

Variant class 3: Video telematics

A dashcam records what is happening in front of the vehicle on a rolling basis. The videos can be analyzed in the event of an accident or for driver training purposes. The systems also work as a tracking computer with documentation of stops, kilometers and routes. Driving behavior (abrupt turns, frequent abrupt braking, etc.) is recorded and sent to the server.

Figure 15Dashcam with location function, driving direction recording and video recording in the event of an accident

Variant class 4: Order management with tablet

Figure 16Truck tablet 6'' with scan option

Instead of the truck tablet in portrait format, tablet computers designed for rugged use are becoming increasingly popular. The driver can easily read all order data on a 8-inch screen, usually in landscape format, make phone calls directly from the order and navigate to the destination. The navigation optionally determines the rolling arrival time for each customer on the tour and transmits this to the server.

At the customer's premises, the driver takes the device out of the holder and generates an electronic signature on a delivery bill PDF, which is then sent to the customer by e-mail. At the customer's premises, he can add photos to the order data in the event of any damage.

Figure 17Truck tablet 8'' with NFC for working time recording

The orders are sent to the truck tablets via the built-in SIM card or via WLAN. If the SIM card is used, the dispatcher already has delivery receipts available during the tour for further processing for invoicing.

Barcode scanner

They are connected to the tablet directly or via Bluetooth. This allows products to be scanned in the loading area and at the customer's premises and compared with the target quantities. This helps to ensure that the delivery is complete.

Rear view camera

Modern systems in this order class can display the image from the vehicle's reversing camera on the tablet using a special holder. This eliminates the need for a reversing camera monitor and there is only one central screen in the cockpit.

Security systems

What new technologies are available to actively warn drivers in the vehicle if they exceed safety limits? Fatigue detection systems react with an alarm sound if the professional driver is frequently inattentive.

New retrofit systems raise the alarm in dangerous driving situations. They cannot be switched off and report unsafe driving directly to the server so that dispatch can intervene with regard to breaks and training measures.

Figure 18Truck tablet 8'' with reversing camera connection and person recognition

Artificial intelligence is also making its way into trucks as part of digitalization. While reversing cameras and side cameras used to only show the image after reverse gear was engaged or the blinker was activated, modern AI cameras (AI = artificial intelligence) with artificial intelligence directly in the camera or a control box make it possible to detect people and forklifts in a previously defined zone and then issue an alarm. The alarm is issued immediately on the tablet within the camera image. Accidents are avoided by the driver reacting quickly.

Today, telematics can therefore not only be seen as a tool for fleet optimization, but also actively contributes to road safety.

Tracking computer

Modern truck tablets can evaluate the data from the tracking computers via a serial cable connection or Bluetooth and then also show the driver technical data such as compartment temperatures, door/flap status, tire pressure, etc. on the truck tablet.

Pay attention to modularity! If you want to start small with variant 1, you should make sure that you can upgrade to variants 2 to 4 later on.

You should be aware that option 1 is inexpensive and simple, but that you do not have to look at the map every day to check on the driver. It is also important to communicate properly with the driver so that they do not feel monitored and the relationship of trust suffers. This will be discussed from page 85 onwards.

This is almost impossible with Variant 4, because here telematics becomes an auxiliary tool for the driver: to achieve a better driving style, for smoother handling, with fewer phone calls, reduced search effort, increased safety, less paper, scanning of documents, avoiding detours, etc.

Drivers like the truck navigation and the integrated reversing camera. The scheduling department has an easier time with order processing and can also see the positions and optionally the arrival times in the event of customer queries.

Before deciding on variant 3, it is important to have a specialist clarify the connection options to the merchandise management system.

Order data must be transferred from the ERP system to the truck tablet. Either there are interfaces, such as with

- Milling industry: VAS Software GmbH,
- Fresh food service: CSB-System SE,
- Beverage wholesalers: ORGA-SOFT, Organization and Software GmbH,
- Brewery: Tobax Software GmbH,
- Textile Service (TIKOS) SoCom Informationssysteme GmbH,
- ERP: Oracle

which use modern JSON/rest protocols and an Oracle cloud data connection to supply the orders. Or, as a minimum, you have an Excel spreadsheet in which the orders are entered and then transferred to an order server. Within 2 to 3 working days, even an exotic ERP system can usually be converted for order data transmission.

With variant 4, you can digitize your workflow in addition to variants 1-3 and at the same time have the costs in relation to the unloading quantity under control for efficiency comparisons. You can see the customer costs per transport unit, the target times for unloading compared to actual times and optimally planned routes compared to actual routes. This variant is more complex, but is mastered by experts, and you benefit from a high level of efficiency with increased driver motivation.

With variant 4, you can also map complex logistical processes - particularly interesting for beverage wholesalers and breweries. The unloading location of pallets on the truck is usually determined by the driver before the new tour with the help of the tablet, as he knows the unloading conditions exactly. However, he is not present during night loading. The tablets from variant 4 can also be mounted on forklifts.

These forklift tablets receive the loading sequences previously predefined by the driver on the truck tablet in the truck, and the forklift driver can then take these into account digitally and load them correctly. The load securing is then also documented on the forklift tablet with a signature.

Dieter Schäfer, a former police director from Mannheim, has been working since 2014 to reduce the risk of accidents involving truck drivers, in particular to avoid death at the end of a traffic jam. He has developed important rules that should be taken into account, especially when using telematics and digitalization in the fleet.

He has also published a book on this subject, which contains the following rules and safety registers:

The 10 max eighty rules for truck drivers

1. Always adhere to driving and break times, use breaks sensibly.
2. Only make urgent calls and only if a hands-free system is available. Every phone call is a distraction.
3. Social media use via smartphone, notebook, tablet, etc., such as Facebook, WhatsApp, SMS or YouTube must be avoided.
4. Do not make arrangements for routes, route changes or order processing during the journey.
5. Driver activities such as reading, making coffee, preparing meals, personal hygiene, etc. are

incompatible with safe driving and must be avoided.

6. Only eat and drink while driving if it is ensured that your concentration on the traffic is not disturbed.

7. Alcohol and other substances that affect reaction and concentration, as well as medication that precludes driving, are strictly prohibited directly before and during driving times.

8. Do not change clothes until the next parking lot or during the break. Wear sturdy shoes.

9. If something falls, stop at the next opportunity and only then pick it up.

10. Max eighty rule: "I adhere correctly to the specified speed limits - especially on routes where there is a risk of traffic jams and at roadworks. I keep safe distances and overtaking bans and I am alert."

(Schäfer 2025)

1.1 Familiarity and knowledge of the assistants. Modern emergency technology is installed in almost every truck. The professional driver must therefore receive instruction on every (!) truck he uses and be familiarized with the available assistance systems. In particular, they are expected to have a sound knowledge of the various override options of the emergency brake assist in the event of danger in order to avoid unintentional deactivation of the emergency braking system in the event of stress. The professional driver must insist on the instruction drive before taking over the truck.

1.2 Adaptive cruise control (ART) Use outside built-up areas is a contractual obligation. The ART leads to a more defensive and quieter and therefore safer drive. It also promotes an economical driving style.

1.3 Truck navigation app Only truck-compatible navigation systems should be used in a truck fleet. The software receives regular updates. The professional driver must familiarize himself with the route and the roadworks situation on unfamiliar routes. Increased attention is expected for traffic jam instructions.

1.4 Deadlines and time slots Delivery and collection deadlines, as well as booked time slots, are binding and must always be adhered to. Time buffers are planned for this purpose. If delays occur that jeopardize deadlines, the customer will be informed immediately.

Making up lost time through speeding or other unauthorized actions is prohibited.

1.5 The 10 Max Eighty rules As a participating logistics or forwarding company, we are committed to compliance with the Max Eighty rules by our drivers. The professional drivers are trained in driver meetings and with the act of a self-commitment to "Max Eighty".

The particular benefits that telematics can offer in terms of safety are described under the heading Accessory connection and video telematics. Telematics not only records, but also sends data to the dispatch workstation, thus creating the basis for training measures for the driver.

Many companies are afraid to make changes to management processes because a shortage of staff on the market could cause drivers and dispatchers to resign, because they feel controlled or because they shy away from the efforts associated with change. This overlooks the fact that with the right communication, many positive effects can be derived from the change.

In the mill, production employees log in to the time recording system in the morning and log out again in the evening, and you know exactly how much flour was produced during the day, which makes it possible to measure performance. The industrial process would be unthinkable without performance measurement and performance-based pay.

Until now, this was not possible in the vehicle fleet and was not practiced. The new technologies make it possible if you use them correctly.

So what needs to be communicated to the driver so that he recognizes the advantage that telematics brings him and sees where the benefit is for the company?

First of all, it is important to bring him to the scheduling workstation and show him that his stops and routes are currently visible in the scheduling system. Standing times with customers and for breaks as well as driving times are recorded and are used, among other things, to determine the costs of the transport and to optimize the routes.

The driver receives an RFID chip to log on in the morning, so it is clear that working time and yard time are recorded digitally.

Good drivers also don't mind having their route visible, as this means they are no longer asked on the phone where they are for so long. Instead, the dispatcher can see that the truck is stuck In a traffic jam, for example.

The tablet: The use of the tablet confronts the driver with a user interface, a computer in the vehicle. A 55-year-old driver recently said: "Not my thing, I've been driving with paper for 30 years". But then his smartphone beeped and he received a WhatsApp message from his daughter. There are a variety of such excuses. It is therefore important to instruct drivers on how to use digital order management on the tablet via video tutorials and individual training. A particularly well-trained driver can also act as a trainer for their colleagues.

The flood of paper is reduced for him thanks to electronic delivery bills and much more. And of great importance: the boss now sees his performance and can reward it accordingly, similar to what has been the case in production for years.

All of this needs to be communicated in the run-up to the introduction of telematics.

Information for the dispatcher

The dispatcher: He knows the customers, is under constant stress and makes sure that the vehicles get on the road. His primary objective is to ensure that the customer is served on time - no matter how - and if the telephone sales department has forgotten to call the customer about the order and then a subsequent delivery is due, it doesn't matter, the main thing is that the customer is satisfied. New telematics solutions show dispatchers the locations of customers on electronic maps and plan routes at the touch of a button. This raises questions about possible threats to jobs in the dispatching department and also about claims that the computer only does "nonsense".

In this context, too, the challenge arises of communicating correctly and making it clear that the computer provides support by showing the predefined regular tours as well as

Figure 19Stops with planned route

which customers have not yet ordered, but should not have any more goods due to the delivery frequency and should be called. It creates a preliminary calculation and helps to identify how transport could be handled more cost-effectively through better capacity utilization.

The dispatcher benefits from a much quieter workplace. The ETA monitor (ETA = Estimated Time of Arrival) enables him to recognize disruptions on the route. He receives an alarm message both on the map and in list form if a planned time window to the customer is not met. This enables him to act in advance and saves him several calls with the constant question about the arrival time of the goods.

He also no longer has to phone the vehicle and be annoyed that the truck is probably in a dead zone or that the driver is not answering the phone so that he does not know where the vehicle is at the moment. Time sheets and trip reports are also available online thanks to the digitalization of the fleet. This allows for more peace of mind in the dispatching department and concentration on optimized planning.

With more information, the position of the dispatcher becomes that of a fleet manager, a helmsman who can make much better decisions based on data than before. They can individually adapt optimized routes in a DispoBoard and enrich them with their knowledge. The job is therefore not under threat, but rather has a significantly higher value for the company, as this is precisely where the efficiency of the fleet is increased in the run-up to the tour.

Information to the controller

The controller: This is a big term and in many companies there is no controller at all, but there is an accountant who is familiar with the accounting data. Controlling means recognizing deviations from target values and taking countermeasures, raising a hand and helping those responsible to do things better. The controller must therefore initiate and moderate control measures using score cards and traffic light colors.

Figure 20Consideration of special unloading conditions when evaluating the customer

The controller or accountant, like the driver and dispatcher, must therefore be involved right at the start of the telematics project, as he is a key figure who provides impetus for more efficient action through the valuable telematics data.

Information to the managing director

The Managing Director: He gives the go-ahead for the telematics project and expects efficiency gains. This does not happen automatically through vehicle tracking. It is important that the management systematically determines the individual work steps from the outset and consciously keeps the objectives and target values in mind. Before starting the telematics project, the actual costs per ton, bag or other transport unit - if possible per truck or tour - must therefore be broken down. The basic data for this is provided by the driver card from the tachograph and a cost and quantity analysis of historical data. This is followed by a monthly comparison with the corresponding actual data, which also reveals where improvements have been made.

It is important that the management is familiar with the key logistics indicators and knows how they are calculated and what information they contain. For example: What influence do the fixed costs, the idle costs, have if the truck stands idle in the yard for a day? How is productivity measured, etc.?

Figure 21Graphical evaluations with score cards

The fleet in your pocket: the managing director should be able to see with just a few clicks on their smartphone or tablet which key figures got out of hand in the past week - but also where improvements have been made that need to be communicated

Important: Action. It must be clearly regulated who has to question or re-regulate what in the event of a deviation from the target and how the deviation from the target is documented. Today, the driver has every opportunity to do this on the tablet.

When the truck leaves the yard, this event is like a black box without the use of telematics. There is uncertainty about arrival times: It is unclear whether customers are assigned to the right tour and the right day of the week. Are the delivery costs for the customer within the agreed and previously calculated framework?

In a realistic example calculation, a detour of 30 kilometers or half an hour of additional downtime has an impact of 12.5% on the cost per ton.

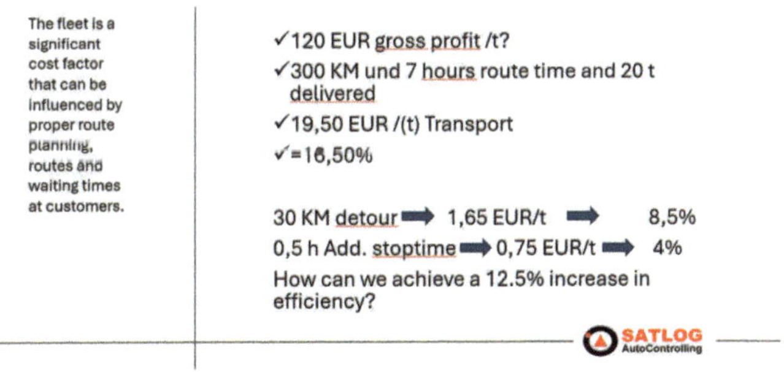

Figure 22Sample calculation of savings effects

But how can you bring transparency into this black box? It's not that difficult. You just have to take a few

hours, calculate the costs, hand over the orders to the route optimization department,
install telematics systems and start taking measurements in order to know more than just making decisions based on gut instinct.

Many of these components are subsidized by 80% for toll vehicles. Not forgetting that there must be a person responsible for projects and services in the company who covers all telematics components.

All in all, the costs of achieving the efficiency targets are generally between 1% and 2% of the annual fleet costs including driver costs.

To be on the safe side, a proof of concept should always be carried out before using telematics. This involves evaluating the tachograph data according to shifts, kilometers, idle times and driving times and comparing the orders of the trucks with the sequence of deliveries carried out in the actual state. The route optimization of these orders then leads to insights into how many kilometers could have been saved by digitizing the fleet and using optimization algorithms. You will be surprised: efficiency increases of between 10 % and 20 % are possible through the optimization and digitalization aspects described in this book, depending on the degree of organization.

THE AUTHOR

Dr. Jürgen Stausberg is author and managing partner of SATLOG GmbH.

He and his team have carried out projects in the following sectors: Bakery, construction logistics, brewery, vehicle transportation, fresh produce service, beverage wholesale, sanitary wholesale, food wholesale, mineral oil trade, dairy, mill, press, recycling, forwarding, textile service and therefore has a very broad knowledge of the industry.

He is an expert in increasing efficiency in the fleets of commercial and industrial companies. The focus here is on cost and performance measurement with the help of telematics and cloud services.

SATLOG was founded in 2000 to optimize the logistics processes of companies with their own vehicle fleets using satellite positions (GPS data).

Dr. Stausberg has dedicated itself to AutoControlling in order to enable the end user to manage and increase the efficiency of his fleet automatically using logistics key figures.

Book author:

The problem of realizing performance-oriented behaviour of drivers in the company's own fleet: ISBN: 3831130299 - German

Fleet controlling with telematics: GPS - GPRS - vehicle location - customer result calculation - logistics key figures (A way to reduce costs in the fleet) ISBN: 9783837018646 - German

Figure 23Dr. Jürgen Stausberg

BIBLIOGRAPHY

Benston, Gorge J. 1972. *The role oft he Firm's Accounting System for Motivation, in: Anton, Hector R. und Firmin, Peter A. Contemporary Issues in Cost Accounting, A Discipline in Transition, 2nd Edition.* New York, Atlanta, Geneva,Dallas, Paolo Alto : Anton, Hector R. und Firmin, Peter A. Contemporary Issues in Cost Accounting, A Discipline in Transition, 2nd Edition.

Drucker, Peter F. 1954. https://www.wirtschaftszitate.de/autor/drucker-peter-f/.

Latham, Gary P. Baldes J. james. 1975. *The Practical Significance of Locke's Theaory of Goal Setting in: Journal of Applied Psychlogy.* Journal of Applied Psychlogy Vol.60.

Lattmann, Charles. 1977. *Führung durch Zielsetzung.* Bern.

Lavinsky, Dave. 2025. *The Two Most Important Quotes In Business.* https://www.growthink.com/content/two-most-important-quotes-business.

Schäfer. 2025. https://road-safety-charter.ec.europa.eu/de/road-safety-in-

action/good-practice/hellwach-mit-80-kmh-ev-
fight-against-distraction-and-microsleep.

Stausberg, Jürgen. 2001. *Das Problem der Realisierung leistungsbezogenen Verhalts des Fahrpersonals im werkseigenen Fuhrpark.* Weinheim: Books on Demand GmbH.

Trolle Lagerros, Ylva. 2009. *Physical activity—the more we measure, the more we know how to measure.* European Journal of Epidemiology | Ausgabe 3.

Wikipedia. 1891.
https://en.wikipedia.org/wiki/Lord_Kelvin.

—. 2025.
https://en.wikipedia.org/wiki/Fleet_Managem
ent_System.

—. 2025. *Digitization.*
https://en.wikipedia.org/wiki/Digitization.

—. 2025. *Route Planning.*
https://en.wikipedia.org/w/index.php?search=
route+planning&title=Special:Search&profile=a
dvanced&fulltext=1&ns0=1.

—. 2025. *Sensors.*
https://en.wikipedia.org/wiki/Sensor.

—. 2025. *Telematics.*
https://en.wikipedia.org/wiki/Telematics.